Biochemistry and daily life.

A person's daily caloric expenditure is determined by factors such as exercise level, body composition (particularly the ratio of fat to muscle), and resting metabolic rate (BMR).

Written by
SARAH OLIVER

Table of Contents

How does it work in your body? 9

The importance of biochemistry derives from its scope; 15

Why do we need biochemists? 15

Add your creative thinking to the technological revolution 15

The process of creating new drugs 17

Clinical Biochemistry 17

Biochemistry is essential to the practice of medicine 19

Explain biochemistry. 21

The science of molecular genealogy 23

What exactly does one do in the field of biochemistry? 25

How has biochemical study influenced medical care? 27

Symptoms of acidosis and alkalosis 30

Where does biochemistry go from here? 32

Conclusion 34

Hormonal imbalance 40

Nutritional Biochemists 43

Endocrinologist 45

Pharmacist 47

anesthesiologist 50

1. Enzymology 53

2. Endocrinology 54

3. Molecular Biology 55

5. Cell Biology 57

6. Structural Biochemistry 58

7. Metabolic Biochemistry 59

8. Immunology 59

9. Neurochemistry 60

Applications of biochemistry in medicine 60

1. Physiology 61

2. Pathology 61

3. Nutritional deficiency 62

4. Hormone deficiency 63

Applications of biochemistry in medical treatment 63

1. Kidney function test 64

2. Blood test 65

3. Liver function tests 65

4. Serum cholesterol test 66

5. Pregnancy test. 66

6. Ames 67

7. Breast Cancer Screening 67

8. PKU 68

Applications of biochemistry in agriculture 68

1. Prevent diseases 69

2. Improve growth 69

3. Improve performance 70

4. Condition of the soils 70

5. Adulteration 71

6. Biochemical tests 71

7. In livestock 72

8. Fishing 72

Application of Biochemistry in Nutrition 73

1. Food containment 73

2. The role of nutrients 74

3. Nutritional value test 74

4. Food limitation 75

Applications of biochemistry in pharmacy 75

1. Drug Constitution 75

2. The half-life test 76

3. Medication storage 76

4. Drug metabolism 77

Applications of biochemistry in plants 77

Photosynthesis 78

Breathing 78

Study of different sugars 79

Plants Secondary Metabolites 79

Other functions 80

Use of biochemistry in forensic criminal investigation 87

APPLICATIONS OF FORENSIC BIOCHEMISTRY .90

Use of biochemistry in textiles 91

Spider silk culture with bacteria 92

Nourish the body with algae 94

Properties of enzymes used in textiles 95

Enzymes used in textile processing 97

enzymatic desire 97

Enzymatic pickling 97

Enzymatic bleaching 98

Biopolished 100

Anatomy of a cell 101

What is a cell? 102

105 cells   104

CHAPTER 1

BIOCHEMISTRY

Biological chemistry is the branch of chemistry that examines the chemical components and activities of living things, including those of plants, animals, and microbes.

How does it work in your body?

Because of its focus on the chemistry of living things, it incorporates the methods of analytical, organic, and physical chemists, as well as those of molecular physiologists. Metabolic processes include the breakdown of substances, usually for energy purposes, and the synthesis of complex compounds that are essential for life.

Enzymes are organic catalysts responsible for catalyzing certain chemical reactions; in turn, enzymes require the genetic machinery of the cell to exist. The study of chemical changes in disease, drug action, and other medical fields, as well as nutrition, genetics, and agriculture, is therefore a natural fit for biochemistry.

The term biochemistry can be interchanged with two others that have been around for a while:

Both physiological and biological chemistry are subfields of chemistry. Molecular biology encompasses the subfield of biochemistry that studies the structure and function of macromolecules such as proteins and nucleic acids. Biochemistry is a relatively new scientific discipline, and the term itself came into use around the year 1900. However, it has much deeper roots; its early history is intertwined with those of physiology and chemistry. Despite these early fundamental discoveries, the rapid progress of biochemistry had to wait until the creation of structured organic chemistry, one of the major achievements of the 19th century. Many thousands of unique chemical components make up a living organism. It is a fundamental challenge in biochemistry to determine the specific chemical reactions that take place in living cells to produce these molecules. The investigation of the cellular mechanisms responsible for the synthesis and degradation of organic chemicals found in living cells was obviously based on a firm understanding of the molecular structure of these substances.

There are not many clear boundaries in science, and the fields of organic and physical chemistry on the one hand, and biochemistry on the other, have traditionally been quite similar. Biochemistry is the application of methodologies and theories of physical and organic chemistry to biological questions. Early efforts in this line of research were hampered by a pervasive fallacy in scientific thought: namely, that the changes that matter undergoes in a living thing are not governed by the same chemical and physical rules that apply to living entities. inanimate The Vitalists held this view; they believed that chemical synthesis could never replicate the natural products created by living things. A mistaken regard for living phenomena hampered development at the same time that the practical requirements of man acted to propel the new science forward. Throughout the nineteenth century, as organic chemistry and physics built up an impressive body of theory, the needs of the medical community, the pharmaceutical industry, and the agricultural community provided a constant impetus for the application of new chemical discoveries. to a wide range of pressing practices. problems. The field of biochemistry investigates phenomena at the molecular level. Investigate cellular processes by looking at molecules such as proteins, lipids, and organelles. It also investigates the mechanisms of cell-to-cell communication, which is essential for processes such as proliferation and disease resistance. Biochemists must know how a molecule's structure is related to its function in order to anticipate how different molecules will react with each other.

Genetics, microbiology, forensic science, plant science, and medicine are just a few of the many fields encompassed by biochemistry.

The importance of biochemistry derives from its scope;

The rapid development of the field during the last century. Being a member of this innovative field of study at this time is a tremendous privilege.

Why do we need biochemists?

To better understand how life works, we need new ideas and experiments to help us understand the relationship between health and disease.

Add your creative thinking to the technological revolution .

Join a team that includes chemists, physicists, doctors and nurses, policymakers, engineers, and many more experts. Among the various places where biochemists can be found working are:

- hospitals
- universities
- Agriculture
- Institutions dedicated to food research
- Education
- Cosmetic products

Investigation of Criminal Activity through Scientific Methods

The process of creating new drugs

Clinical Biochemistry

The duties of a clinical biochemist vary greatly by field. Clinical Biochemists provide guidance on the conduct and interpretation of laboratory investigations in non-teaching hospitals as well as in state and private laboratories dealing with a wide range of sample sources (especially general practitioners). However, the general clinical biochemist is increasingly redundant in university hospitals due to increasing laboratory orientation and sub specialization within clinical units, allowing clinical and scientific staff in such units to provide their own core areas of competence. in clinical biochemistry. A clinical biochemist, with either a scientific or medical background, can make a significant contribution to patient care when assigned to a position that involves a formal and joint attachment to a clinical service (eg, a lipid unit, endocrinology or renal) . This position also offers increased job satisfaction and more research and development opportunities at both the clinical and laboratory levels.

The obvious educational and controlling role of the clinical biochemist in this exercise, regardless of primary field of interest, makes cost-benefit analysis in laboratory research a timely and infinitely valuable study.

The role of biochemistry in medicine

The field of biochemistry is rapidly developing and rapidly gaining prominence as one of the most important fields in all of science. Because it integrates the fundamentals of biology and chemistry, it plays an important role in the discovery of new innovative techniques in the scientific community. But what exactly is its importance in the field of medicine?

Biochemistry is essential to the practice of medicine .

We wouldn't have the chemical understanding to build the essential pharmaceuticals, cures, and diagnostic tools used daily if biochemists weren't constantly making advances in their field.

Biochemistry, like the field of medicine, is huge and each year brings new and exciting discoveries. We gain a deeper understanding of the underlying transitions between chemical structures and processes that contribute to human health and disease through the application of biochemical techniques.

The ramifications of learning what goes wrong at the cellular level in disease are enormous. Anyone working in the medical sciences has the opportunity to revolutionize healthcare around the world by leveraging their expertise in biochemistry and related fields such as molecular biology and immunology.

There has never been a greater need for the work of biochemists than now, with the proliferation of public health problems such as air pollution, climate change, NCDs, AMR and dengue.

For this essay, we will focus on the specific reasons why biochemistry is so important to the life sciences and, in particular, to medicine.

Explain biochemistry.

Biochemistry is a scientific discipline that bridges the gap between biology and chemistry, as its name suggests. In the same way that chemistry studies inanimate things, biochemistry investigates living organisms.

Biochemists are scientists who specialize in the study of molecular-level biochemical events that take place in living organisms. Molecular biology and biochemistry are fields that study the human body and how it works, but medical biochemistry focuses specifically on the relationship between the two. The following are all within the realm of medical biochemistry, however these are the most common.

carbohydrates and lipids; amino acids and proteins; blood and plasma; biological membranes; nucleic acids; (DNA and RNA)

Human development, enzyme activity, membrane transport mechanisms, homeostasis, blood coagulation, oxygen transfer, neurotransmitter function, and aging are examples of key chemical processes in the body.

Vitamins and their function in the nutritional and mineral processes of the organism.

The science of molecular genealogy
Inheritance

genomics

Biological macromolecules, the huge, complex molecules (such as proteins) that create the structure of cells and carry out many of the processes associated with life, are the focus of much biochemical research.

The reactions of ions and smaller molecules are also crucial to cellular chemistry. They can be organic (such as amino acids) or inorganic (such as the elements used to form proteins) (for example, water and metal ions).

Biochemistry has also had applications outside of medicine. Cell biology, physiology, pathology, pharmacology, microbiology, immunology, nutrition, forensics, and toxicology are just a few of the areas that have benefited greatly from the study of this topic.

Biochemical studies also have practical applications outside the field of medicine, including in manufacturing, agriculture, and food processing. When these various contexts are considered, it becomes abundantly evident that biochemistry is not a single, unified field.

What exactly does one do in the field of biochemistry?

Biochemists often spend their days running experiments, supervising lab workers, writing their findings in technical papers, and presenting them to colleagues and relevant parties.

Researchers analyze enzymes, DNA, and other compounds using electron microscopes, lasers, and other state-of-the-art laboratory equipment to carry out their investigations.

From obtaining plant and animal cell samples for genetic studies to creating successful new drugs for the pharmaceutical industry, these laboratory experiments cover a wide range of topics. Results are analyzed in an office environment using state-of-the-art data modeling tools once collection is complete.

Biochemists are in high demand in the life sciences industry, where they typically work in research teams alongside experts from other scientific fields (including pharmaceuticals, biotechnology, toxicology, food technology, and vaccine production). Additionally, they are widely used in investigative capacities by public and private organizations .

How has biochemical study influenced medical care?

Ever since Eduard Buchner's discovery in 1897 (widely regarded as the birth of biochemistry) that a cell-free yeast extract can ferment sugar, biochemistry has had a close relationship with medicine, shedding light on many previously obscure elements of medicine. human health and disease.

The metabolism, function, and development of a normally developing human organism are complex subjects, and anyone working in medicine or a related subject needs a solid background in biochemistry to understand even the most basic concepts.

Biochemistry has greatly assisted physiology in its study of the relationship between biochemical changes and physiological states. It is useful for discovering how digestion, hormones, and muscle contractions and relaxations work chemically.

Physicians can use biochemical analysis to corroborate patient-reported diagnoses in the field of pathology, which investigates the relationship between abnormal biochemistry and pathological states of the human body.

If a patient experiences sudden, sharp pain in one or more joints, the doctor may suspect gout, a form of arthritis caused by excess uric acid in the bloodstream. Biochemistry can confirm if gout is the cause by evaluating uric acid levels.

Biochemistry allows us to understand the chemical processes behind diseases as diverse as:

- diabetes
- hyperthyroidism,
- hypothyroidism,
- hyperammonemia .
- low and high parathyroid function conditions
- jaundice
- renal disease
- hypercholesterolemia
- phenylketonuria
- red blood cell disorder caused by the formation of sickle cells

- tooth fluorosis
- rickets

Symptoms of acidosis and alkalosis

diseases caused by a buildup of waste products in lysosomes

atherosclerosis

The job of biochemists in medicine is to study possible therapies for diseases by deducing their chemical composition.

There is usually a correlation between the effect of a drug and a change in the body's biochemistry.

Therefore, pharmacologists must have some understanding of how the human body works biochemically.

Biochemical tests are crucial in the pharmaceutical industry as they reveal information about the characteristics of a drug,

- modus operandi
- half life
- environments to store
- metabolism
- negative or harmful results that could occur

The only discipline that adequately explains how vitamins work in the body is biochemistry. The field of nutritional deficiency will continue to be deeply affected by the continuing discoveries made by biochemists, as millions of people around the world take vitamin and mineral supplements on a regular basis.

In general, the true impact of biochemistry is difficult to assess. The pioneering work of biochemists continues to broaden the horizons of medical science, from lab-grown placentas that will "disrupt pregnancy studies" to novel drugs that eliminate antibiotic-resistant germs.

Is there a way that biology can help curb the overuse of antibiotics? Here you can learn more about this topic.

Where does biochemistry go from here?
The various branches of biochemistry are developing rapidly.

We explore how genomics, the study of an organism's entire DNA pool, is a branch of biochemistry with far-reaching consequences for the pharmaceutical industry and health care systems.

For example, genome sequencing is leading to a "revolution" in early diagnosis, which is a major step forward that could accelerate the introduction of new life-saving drugs for a wide variety of medical problems.

efficiently and effectively analyze huge data sets, allowing them to map biochemistry in ways that have never been done before.

In light of biochemistry's prominent place in the life sciences as a sustainable approach to solving a wide range of scientific, medical, and industrial challenges, the field is poised to experience a dramatic increase in job openings over the next decade. .

Browse the latest biochemistry job openings and take the next step in your professional life.

conclusion

It goes without saying that the purpose of medical science is not limited to elucidating the mysteries of life. It was created for one reason: to make a real difference in the way medicine is practiced today.

This is aided by molecular phenomena uncovered by biochemical analysis, providing research that offers local health professionals more leeway to fulfill their duty of care.

Indeed, the laboratory discoveries made by biochemists will determine the course of medical science for the next decade. Just in the week before this blog was launched, two independent biochemical studies indicated that women have stronger episodic memory than men and that a common acne treatment helps prevent hardening of the arteries. There seems to be an almost infinite space for study.

On Earth, all life is based on metabolic processes and reactions. Biochemistry is and will continue to be one of the most important scientific disciplines because of the way it unites theoretical knowledge with real-world applications in health promotion, disease diagnosis, and treatment development, as well as in our quest to understand the origins of life on Earth. Understanding and maintaining health, and understanding and effectively treating disease, are the two main interests of professionals in the health sciences, especially physicians. Biochemistry has an effect on these two central themes, and the interaction between biochemistry and medicine is expansive and mutually beneficial. Biochemical research has shed light on previously obscure issues related to health and disease, and vice versa. Analysis of various variants of sickle cell and other hemoglobins has contributed considerably to our understanding of the structure and function of normal and sickle cell hemoglobin , as well as other proteins. In the early 20th century, the English physician Archibald Garrod investigated individuals with the unusual diseases of alkaptonuria , albinism, cystinuria , and pentosuria and found that these conditions were dictated by genetics. Garrod classified these disorders as inherited metabolic disorders. The advancement of human biochemical genetics can be traced back to his observations. Examining the causes of familial hypercholesterolemia, a condition that leads to premature atherosclerosis, is a more contemporary example. This not only helped shed light on the various genetic mutations that cause this disease, but also shed light on the receptors on cells and the mechanisms by which chemicals like cholesterol are taken up by them. As a result of the investigation of oncogenes and tumor suppressor genes in cancer cells, scientists have begun to focus on the molecular mechanisms that regulate normal cell proliferation. These cases show how understanding disease can lead to new frontiers in fundamental biochemical research. The scientific method provides a framework for health professionals and biologists that informs their work, sparks their interest, and encourages them to pursue continuing education. As long as medical practice is based on a solid understanding of biochemistry and other basic sciences, it will be able to rationally incorporate and adapt to new information. Clinically significant function:

Without the study of biochemistry, medical science would not have progressed as far as it has.

understanding of the related biochemical and physiological changes that occur in the body in response to infection or disease.

The patient's history can give the doctor clues about metabolic abnormalities and underlying diseases. A physician may suspect gout if a patient has complaints of multiple minor joint stiffness, and this suspicion could be validated by measuring uric acid levels in the blood, as gout is caused by the buildup of uric acid in the blood.

hormonal imbalance

It is one of the main causes of disease, especially in women and children. Biochemistry classes help doctors understand their patients' symptoms by explaining the importance of hormones in maintaining health. Professional in Clinical Biochemistry

The Clinical Biochemistry of Laboratory Medicine subspecialty deals with the analysis of patient samples and their interpretation by physicians. A clinical biochemist is a medical expert who works in a laboratory and analyzes body fluids, tissues, and other body substances for markers of disease.

In the absence of detectable chemical or biological abnormalities, diagnostic approaches used in analysis are often automated procedures. Electrophoresis, electron microscopes, spectroscopes, and chromatographs are just some of the equipment clinical biochemists use to perform advanced tests as soon as abnormalities are discovered.

In addition, clinical biochemists must analyze the data and communicate their findings to the appropriate parties. The decision to become a clinical biochemist is fraught with great responsibility, as the success of diagnosing and treating a patient depends on the precision and accuracy of the results they produce. Therefore, clinical biochemists need not only a strong academic foundation in the subject matter of competency, but also additional skills and attributes, such as drive, focus, dedication to patient care, independence, and the ability to work in a team. A Nutritional Biochemist . Nutrition is the science that provides detailed information on the interaction of nutrients and other substances in food with respect to the development, growth, reproduction, health and disease of a living organism. Nutritional biochemistry, on the other hand, integrates biochemistry and nutrition to examine and describe the role of dietary requirements for human health at the cellular and molecular levels. More specifically, nutritional biochemistry involves concepts, ideas, and methodologies related to the chemical properties, biochemical, metabolic, and physiological functions of various nutrients and dietary constituents.

Nutritional Biochemicals

are generally responsible for determining optimal dietary intakes for each nutrient and food component throughout the life cycle of an organism.

A nutritional biochemist is an expert in the field of food and nutrition who advises people on how to take dietary supplements to maintain a healthy lifestyle or improve health-related problems. One of several workplaces for nutritional biochemists is a hospital where they assess their patients' health conditions based on which they propose individual treatment plans.

Because nutrition is a crucial aspect of a healthy lifestyle, nutritional diseases can cause disease in humans. Such diseases can range from deficiencies or excesses in an individual's diet to obesity, eating disorders, and chronic diseases (eg, diabetes, hypertension, cancer, cardiovascular disease, etc.). Due to this, Nutritional Biochemists occupy an important place as professionals in various hospitals or health centers.

endocrinologist

Endocrinology is the field of science that studies the endocrine system in the human body. The endocrine system is a system of glands that are responsible for the secretion of hormones. Additionally, Endocrinology explores a wide variety of hormone-related diseases that affect the proper functioning of the body's different organ systems. Consequently, an endocrinologist is a doctor specializing in the endocrine system who can diagnose and treat various diseases associated with hormones. Although some disorders are related only to the endocrine system and can only be detected by endocrinologists, there are many ailments that can have not only endocrine but other origins as well. In such cases, endocrinologists work in collaboration with other professionals to identify and propose effective treatment.

To become an endocrinologist, individuals must first earn a bachelor's degree in medicine, chemistry, biology, biochemistry, or a related discipline. After graduating from college/university, people go to medical school for another 4 years and work in hospitals and clinics as residents for 3 years. After the last 2-3 years of training specifically in endocrinology, they are eligible to work as endocrinologists in hospitals.

Pharmacist

Pharmacy is a science that studies the preparation techniques, uses, effects and mechanisms of action of various drugs. Furthermore, Pharmacy deals not only with the chemical properties of medicines, but also with their impact on living organisms.

On the other hand, a Pharmacist is a health professional specialized in Pharmacy, focusing on the safe and effective use of pharmaceutical drugs. Pharmacists not only work in pharmacies, but also in various health care facilities, including hospitals. Such pharmacists are called hospital pharmacists.

Hospital pharmacists are responsible for determining the optimal form of medication for each patient. In addition, your duties also include resolving patients' medical condition and medication history, screening a patient for any drug allergies, verifying a patient's valid medication history, reviewing and ensuring the accuracy of prescriptions of doctors, keeping records of different pharmacies files, follow-up of medication charts, preparation and discharge of patients.

Along with the work that Hospital Pharmacists perform independently, there are issues that require the cooperative work of Hospital Pharmacists and hospital staff members (eg, dieticians, physicians, etc.).

Working as a Hospital Pharmacist is a very responsible profession since each and every one of the decisions must be made efficiently and in the shortest possible time considering several aspects that were mentioned above. Since Hospital Pharmacists play an essential role in the proper treatment of patients, becoming a Hospital Pharmacist requires a full qualification in the field of specialization along with the necessary experience in the area.

anesthesiologist

Anesthesiology is the branch of medicine that deals with the complete care of the patient before, during and after surgery. Anesthesiology involves the use of various injected or inhaled medications to render the patient numb. A doctor who specializes in this field is known as an anesthesiologist.

An anesthesiologist is responsible for the proper implementation of anesthesia and anesthetics to safely support the patient's essential functions during the perioperative period. The duties of an anesthesiologist include evaluating the patient before surgery, consulting with the surgical team, managing pain and monitoring vital functions during the operation, monitoring the conditions of patients after surgery, and discharge of patients.

Considering the fact that there are no specific anesthesiologist specializations, individuals wishing to become anesthesiologists must obtain a bachelor's degree in biology, chemistry, biochemistry, health sciences, or other medical-related disciplines and complete additional specific training in anesthesiology.

Being an anesthesiologist requires in-depth knowledge of organ support techniques through which they assess a patient's medical readiness for surgery. The anesthesiologist plays one of the most critical roles within hospital facilities since without him it would not have been possible to perform various necessary operations and procedures.

Main Branches of Biochemistry

The sub-study of biochemistry is quite diverse as it will follow the progress of other studies such as biology, chemistry, and physics. Previously, the study of biochemistry mainly talks about the behavior of proteins and enzymes. Today, there are many branches of biochemistry dealing with different types of components such as DNA and RNA research, protein synthesis process, cell membrane, and many more.

So here are some popular branches of biochemistry among many other branches.

1. Enzymology

This is the study on enzymes. This type of study will cover the properties of enzymes or biological catalysts. This study includes certain proteins, certain catalytic RNA, and coenzymes and cofactors such as metals and vitamins. In this study of enzymology, you will get the definition of catalysis, the process of enzymes: the catalytic transition stage of the substrate, activities of enzymes, reaction kinetics, enzyme regulation from the biochemical perspective. You will understand about the catalytic effects that occur in these biological elements.

2. Endocrinology

In short, endocrinology is the study of hormones. Hormones are the internal secretions produced by the special cell with the ability to affect the functions of the other cell. This study also covers biosynthesis, function and storage of hormones, cells and tissues, hormone signaling process and many other hormone related processes in living organisms. In this study, you can also learn the medical endocrinology, plant endocrinology, and animal endocrinology.

3. Molecular biology

The study of molecular biology is the discipline that aims to understand the chemical process that occurs in the living organism from the molecular point of view. In this study, you will cover the detailed information about classical biochemical and metabolic cycle. You will also learn about the integration and disintegration of molecules in the living organism. You will see the properties of biological macromolecules such as enzymes, DNA, RNA, hormones, etc. in the living cell. You will understand them and know exactly how they work at the molecular level.

4. Molecular Genetics and Genetic Engineering

This study is the mix between biochemistry and molecular biology that covers genes, their inheritance and expression in more detail. In this study, you will learn deeply about DNA and RNA, their special function tools like massive sequencers, PCR, translation, DNA and RNA extraction process, in vitro and in vivo mechanism, enzyme restriction, DNA replication and much more. plus. If you really want to understand about DNA and RNA then this study is right for you. In the study of molecular genetics and genetic engineering, you will also learn about genes, how they insert, how to silence them, and their special expressions and properties. This study aims to break the limits in the species since the genome of the species can be inserted from one species to another species. This study also aims to create the proper regulation on gene expression.

5. Cell Biology

This study gives us information about the properties, physical structure, biochemical composition, functions, life cycles of cells. The life process of cells includes their nutrition, respiration process, defense process, their division, etc. This study also covers the communication that takes place between cells in the multicellular organism. There are also some techniques used in this study consisting of cytochemical techniques, plant cell cultures, light and electron microscopy observation, ELISA or flow cytometry and many more. The study of cell biology has a strong correlation with histology, microbiology, and physiology.

6. Structural Biochemistry

This study aims to provide clear information on the biological architecture of macromolecules such as proteins and nucleic acids (DNA and RNA). This study covers peptide sequences and the physical-chemical atomic interactions that allow these structures. There are some current studies on protein structure that use the basic concept of this structural biochemistry study.

7. Metabolic Biochemistry

This metabolic biochemistry study will help you understand the different types of metabolic pathways at the cellular level. This study can give us detailed information about the biochemical reaction in the cell and allow us to understand the metabolic disease. This study can be used to make a great treatment and medicine for metabolism related disorders.

8. Immunology

This study will cover chemical reactions and the function of the immune system in the living organism. Bacteria, viruses and their chemical reactions will be discussed. This is the study that gives the great contribution in the understanding of antibodies.

9. Neurochemistry

This study talks about neural activities within organic molecules. It encompasses neurotransmitters and other molecules, such as neuroactive drugs that influence neuronal function.

Meanwhile, here are the applications of biochemistry in some industries:

Applications of biochemistry in medicine

The study of biochemistry is actually related to the medicine industry. The concept of biochemical study has been widely used for the improvement of medicine. So here are some great applications of biochemistry in the medical field.

1. Physiology

Biochemistry is really helpful for scientists to fully understand the physiological of the human body. By using the concept of biochemistry, scientists can understand the biochemical changes that occur in people. The search for the cure of diseases based on biochemical changes has been carried out to obtain the best medicine for certain diseases.

2. Pathology

Biochemistry is also used in pathology. By using this study, the doctor can get clues about the biochemical changes in the patients' bodies based on the description of their symptoms. The doctor can then confirm the patients' symptoms by examining the chemical reaction or process in their body. For example, the doctor may examine the level of uric acid in the patient's body to understand the patient's symptoms.

3. Nutritional deficiency

As we know, it is very important to have a healthy life. Today, people consume multivitamins and minerals to stay healthy. To formulate the best composition of these multivitamins or minerals, scientists use the discipline of biochemistry. The function of the vitamin or mineral in our body can be understood through the study of biochemistry.

4. Hormone deficiency

Hormonal imbalance, whether in men, women or children, can lead to many diseases. In correlation with the study of biochemistry, doctors use the concept of biochemistry to deeply understand hormones, their functions and reactions in the human body. By using this, the doctor can create the right medication to treat that type of hormonal imbalance.

Applications of biochemistry in medical treatment

Biochemistry also has an important role in the nursing discipline. Biochemistry can be used to monitor the progress of patients regularly in the hospital. This study can be used to define the best treatment for patients based on their conditions. Biochemistry makes it easier for nurses to diagnose the clinical status of patients. Here are some other possible applications of biochemistry in nursing.

1. Kidney function test

Renal function is closely related to the biochemical study. If you have kidney problems, for example, you can undergo the urine test. This urine test can help us understand the PH change, urine color or other urine retention. Biochemistry is the fundamental study that covers this type of treatment.

2. Blood tests

The blood test uses the concept of the biochemical study. Analytical test for blood glucose level gives insight about the patient's condition. For example, for the patient with diabetes, this type of test can give an indication about the stages of diabetes. There is also another test for diabetes patients called a urine ketone test. Ketone bodies or ketone urea is the indication of the last stage of diabetes. The containment of this substance in our urine can be examined by studying biochemistry.

3. Liver function tests

Another organ of the body that can be understood by biochemical study is the liver. Biochemistry helps us understand how the liver works, the type of disorder that is likely to attack the liver, and the effect of certain medications on liver disease.

4. Serum cholesterol test

Biochemistry can also be used to check your blood cholesterol level. The cholesterol or other lipoproteins in your blood can indicate your health status.

5. Pregnancy test.

The pregnancy test uses the presence of HCG or the hormone human chorionic gonadotropin as an indication of pregnancy. This type of hormone is produced by the placenta as soon as the embryo implants itself in the uterine walls. This hormone can be detected by urine or blood sample. Thanks to biochemistry for helping us discover this hormone.

6. Ames test

The Ames test is the test that is applied to salmonella bacteria. This test measures the growth of this bacteria in the body. This Ames test helps us examine chemicals to understand if these bacteria mutate the DNA structure. This test can also give us the indication of the cancer potential in our body based on that chemical reaction.

7. Breast Cancer Screening

Screening for breast cancer is done by examining mutations in two genes consisting of breast cancer gene-1 (BRCA1) and breast cancer gene-2

(BRCA2). The examination of these genes can be done with the help of the biochemical concept.

8. Phenylketonuria test

PKU or phenylketonuria is a metabolic disease in which the individual lacks an enzyme called phenylalanine hydroxylase. The absence of this enzyme can generate a phenylalanine, which can cause mental retardation.

Applications of biochemistry in agriculture

Biochemistry also has a valuable impact in the fields of agriculture. This study is useful in agriculture, fishing, poultry farming, sericulture, beekeeping and many more. So, here are some popular applications of biochemistry in agriculture.

1. Prevent diseases
The concept of biochemistry can be used to create the best method of disease prevention. It can also help understand the treatment of certain diseases.

2. Improve growth
Biochemistry can help the farmer improve their plant growth, yield and feed quality by maximizing their fertilizers. Biochemistry helps us understand the reaction that occurs between fertilizers or pesticides and crops at the cellular level. Knowing this, we can choose the right fertilizer

to improve the growth of our plants and their quality.

3. Improve performance

There are a few different hormones that are used for different purposes. For example, there is the hormone that helps growth, or stimulates flowering, or supports the formation of fruits. In fishing, hormones are used to stimulate the growth of fish. These can only be achieved with the help of biochemical study.

4. Soil condition

Understanding the condition of the soils is really important to make sure we get great farming results. Biochemistry can help farmers understand the composition of soils and their deficiencies. These data can be used to define the appropriate actions to solve problems in agricultural soils.

5. Adulteration

Adulteration is considered a serious problem in the field of agriculture. Biochemistry can help us understand the composition of food and its alterations. Biochemical testing can be used to prevent contamination in the food product.

6. Biochemical tests

Biochemical tests for pesticide residues or other toxic wastes in plants, food grains, and soil may also be performed. This can be useful in the import and export process so that the quality of the food is well maintained.

7. In livestock

Biochemical test can be done to check the quality. It can also help diagnose any disease in animals.

8. On the fishing

Biochemical testing is used in fisheries to monitor water quality on a regular basis. The composition of the water and its chemical reactions can change and this change can lead to the death of fish or shrimp. In this condition, regular control by biochemical test is a mandatory action to be taken. There are some qualities that are examined in the biochemical test of water, such as salt or calcium content, pH, accumulation of waste, etc.

Application of Biochemistry in Nutrition

Biochemistry also plays an important role in nutrition. Biochemistry helps us understand chemical containment in foods. This can be used to keep us healthy, define the optimal intake of micro and macronutrients, vitamins, minerals, essential fatty acids, water and all important nutrients. So here is a good example of application of biochemistry in nutrients.

1. Food containment

We can understand the containment of the foods we consume such as carbohydrates, proteins, fats and others. We can also understand the possible physiological alteration due to the deficiency of a certain nutrient.

2. The role of nutrients

Biochemistry makes us understand the role of nutrients for human health. We can know how vitamins, minerals and essential fatty acids affect our body. Based on its chemical content, doctors can also recommend the great substance for our body, such as essential amino acids, cod liver oil, salmon oil, and many more.

3. Nutrient Values Test

By using the biochemical test, we can define the percentage or values of nutrients in certain foods that we consume.

4. Food limitation

By the concept of biochemistry, the doctor can prescribe to limit the use of certain foods such as excess sugar for diabetics, excess oil for patients prone to heart and lung problems, etc. We know that these carbohydrate and fat diets can inhibit the rate of recovery from this disorder. .

Applications of Biochemistry in Pharmacy

Biochemistry can also be beneficial in the pharmaceutical industry. In this industry, many drugs are often stored. These are the following applications of biochemistry for pharmacy.

1. Drug Constitution

Biochemistry helps us understand the condition of certain drugs such as their ability to degrade due to different temperatures. This study can help us increase the effectiveness of medications and minimize side effects.

2. The half-life test

This biochemical test is the test performed on biochemical drugs to determine the longevity of the drug. We can understand the stability of the drug if it is kept under certain temperature conditions. Through this test we can obtain the expiration date of the drugs.

3. Drug storage

Biochemical tests can be performed to determine the storage status of drugs. We know that drugs or other substances like enzymes and hormones are stored for dispensing. It may deteriorate due to temperature, contamination, or poor storage conditions.

4. Drug metabolism

Biochemistry helps us understand the process of metabolism in drug molecules, as many biochemical reactions occur in the presence of enzymes. Knowing this, we can understand the side effect of certain drugs for certain patients.

Applications of biochemistry in plants

Biochemistry is also important for plant life. We can understand the chemical reactions that occur in the plant and how we can maximize them to increase our productivity. So, here are some popular biochemical applications for plants.

Photosynthesis

Photosynthesis is one of the chemical reactions that occur in the plant. By understanding this chemical process, we can gain insight into how carbohydrates are made with the help of sunlight, CO2, and water. There are also complex enzymes that help plants in this process. Understanding the basic concept of this chemical reaction can give us an idea of how to improve the quality of the photosynthesis process.

Breathing

Respiration is the chemical reaction that is related to the process of photosynthesis. As a product of photosynthesis, the plant will release oxygen while taking carbon dioxide from the air.

Study of different sugars

Due to biochemistry, we can also understand that there are some types of carbohydrates that are formed in plants. There are trioses (3-carbon sugars like glyceraldehyde), tetroses (4), pentoses (5), hexoses (6 = glucose), heptuloses (7), and so on.

Secondary Plant Metabolites

Biochemistry gives us information about the process of formation of plant products such as tannins, resins, alkaloids, gums, enzymes and phytohormones.

Other functions.

Biochemistry also gives us useful information, such as how the fruits of plants ripen, how the seeds of plants germinate, the process of respiration, etc.

In fact, there are many applications of biochemistry in our daily lives. Furthermore, the branches of biochemistry have been used to make up many industries, be it medical, nursing, agricultural fields, and more.

Increasingly, the global food system is under pressure, with an increase in the prevalence of polarizing obesity and poverty, and an increased reliance on chemical fertilizers and pesticides, poor quality food, environmental degradation and biodiversity loss. As such, many practices are being reviewed and regenerated. These practices are informed by biochemistry.

Biochemistry is used to improve plant growth, yield and quality as a result of optimizing fertilizer components. Crop improvement has also been enhanced by increased tolerance to biotic and abiotic stress, along with increased nutritional value.

With knowledge of the mechanism of action of fertilizers, such as nitrates, the use of fertilizers can be optimized to improve the quality of plant growth. An example of this is the increasing use of biochemical fertilizers including solutions of nitrogen, phosphorus, potassium, sulfur solubilizers, and various fungi such as mycorrhizae and Trichoderma , as well as small molecular weight iron chelators called siderophores that are produced by microbes.

This is believed to enhance the effect of heavy use of chemical fertilizers, which cause water pollution, nutrient depletion and soul deterioration.

Biochemistry plays an important role in nutrition and health and is considered a powerful unsustainable tool for improving health, reducing poverty and hunger in the world. Through the use of sustainable biochemistry, the commercialization of biochemical techniques is seen as a powerful way to reduce global poverty and hunger and improve nutritional delivery worldwide. Applications of Biochemistry in the Fashion Industry

Biochemistry is used in biotechnological applications in the textile industry. Enzymes are commonly used to bleach and wash textiles, and to change the properties of clothing, for example by changing the appearance of denim or preventing shrinkage of fiber types such as wool and cotton. Increasingly, microbial involvement in the fashion industry has begun to take hold, avoiding the use of traditional chemical processes that are associated with high levels of contamination.

Spider silk, for example, stands out for its strong flexible and lightweight properties; however, in the past it has not been possible to grow spider silk on an industrial scale. However, through the use of fermentation bioreactors, genetically modified bacteria can be used to produce this in large quantities. Due to the knowledge of the material properties of silk at the molecular level, this allows significant control over the final product relative to traditional materials. Additionally, this use of technology addresses the issue of sustainability, as silk is produced in the absence of animal or petroleum-derived material. Indeed, biochemically mediated approaches have the potential to affect climate change, which is increasingly recognized as a major challenge facing society around the world. Biochemical knowledge has been used to identify solutions such as algal biofuels, carbon sequestration, and more efficient industrial processes, which can help protect the environment and improve economic opportunity.

Biochemical research can also be used to understand basic biological processes, as well as complex and elegant mechanisms for harnessing energy and converting it into a usable form. By understanding these processes, the development of advanced biotechnological products has been achieved, which allow the production of new types of bioenergy, such as biochemical photovoltaics.

Through the identification of natural products that are produced from biochemical reactions, products that improve human health have been developed. This research has been fundamental and has increased public understanding of the importance of good nutrition and disease.

This article addresses some examples of biochemistry in everyday life. Biochemistry continues to address the challenges facing society around the world, improving and influencing aspects of our lives.

Use of biochemistry in forensic criminal investigation

Forensics is an essential tool in the criminal justice system, particularly when examining physical evidence to support criminal investigations and subsequent prosecutions.

A sub-area of forensic analysis based on biochemistry/molecular biology, forensic serology deals with the complex task of collecting information on the type of sample, age, origin or sex of biological fluids found at crime scenes. Forensic serology is based on two methods: immunoassays and

DNA/RNA analysis. This chapter reviews the biochemical analysis of biomarkers, and these methods are exemplified by analyzing the ethnicity and gender of criminal suspects, as well as determining the age of the blood sample. Knowing the difference in blood creatine kinase (CK) and lactate dehydrogenase (LDH) concentrations in healthy adults from two ethnic groups, Caucasian (CA) and African American (AA), and taking into account the distribution pattern, the authors mimicked samples of different ethnic origins with appropriate concentrations of CK/LDH. First of all, forensic science can be defined as the investigation of crime using scientific techniques and methods. These techniques are used to examine materials that were present at the crime scene. Forensic scientists conduct a thorough analysis of all these materials to obtain clues as to who may have committed the crime. Investigations surrounding murder, rape, and assault rely heavily on the work of forensic science laboratories to point criminal investigators in the right direction.

Forensic investigations often involve serological and biochemical techniques. The biomedical techniques used to conduct forensic investigations constitute the field of forensic biochemistry, which has various applications. For example, forensic biochemists may be asked to trace the origin of a particular substance, determine parentage or relationships shared by specific people or animals, or even trace the spread of disease.

Forensic biochemistry has proven invaluable in conducting forensic scientific investigations, particularly the DNA fingerprinting technique. However, it should be noted that forensic biochemistry should be used with caution, as its findings may have serious implications.

APPLICATIONS OF FORENSIC BIOCHEMISTRY

Some of the ways forensic biochemistry is used include:

Analysis of the evidence found at the crime scene, using biology, chemistry, physics and genetics.

Qualitative analysis of the evidence through punctual tests and microscopy.

Study of body fluids by separation analysis and optical methods.

DNA tests to find out the relationships between two humans or animals.

Trace the origin of specific materials or substances using chemical and biochemical techniques

Use of biochemistry in textiles.

Biotechnology already plays an important role in the textile industry.

Enzymes are commonly used to wash and bleach textiles, to give jeans a denim look, or to prevent wool from shrinking. A new wave of technology could take this one step further. In the not too distant future, our clothes will be made and dyed by live microbes, abandoning many of the chemical processes that make fashion one of the most polluting industries in the world.

Many players around the world are already revolutionizing the fashion industry with biomanufacturing.

Spider silk culture with bacteria

Spider silk is known to be a strong, flexible and lightweight material, but it is not possible to farm spiders on an industrial scale. Based in Germany,

AMSilk uses genetically modified bacteria to fix that problem. Inside fermentation bioreactors, bacteria produce spider silk protein, which is then spun into fibers, creating an entirely new material with unique properties.

"In textiles, we haven't seen a truly new material in decades," CEO Jens Klein told me. AMSilk has just celebrated the launch of the first product made from this spider silk: a luxury watch strap. The company is working on several other products using spider silk fibers, including biodegradable sneakers for Adidas.

Nourish the body with algae

Not only the fashion industry is responsible for 20% of the world's water pollution. The chemicals used to make and dye fabrics are often toxic, injuring and killing thousands of workers. These chemicals can also seep onto the wearer's skin. Enzymes, such as cellulases, catalase, and laccase, are commonly used in the textile industry. These enzymes are used to remove starch, degrade excess hydrogen peroxide, bleach textiles, and degrade lignin. Due to the highly specific, efficient, non-toxic and environmentally friendly characteristics, the use of enzymes in the textile industry is growing rapidly. The application of cellulases for denim finishing and lactases for textile effluent decolorization and textile bleaching are the most recent commercial developments. Additionally, the use of enzymes results in reduced process times, energy and water savings, improved product quality, and potential process integration.

Properties of enzymes used in textiles

First, the enzyme speeds up the reaction by lowering the activation energy and remains intact at the end of the reaction by acting as a catalyst. Second, enzymes operate under milder conditions. The enzymes can be used in catalytic concentrations at low temperatures and near neutral pH values. Third, enzymes are the best alternative to toxic, dangerous and polluting chemicals. Fourth, the enzymes act only on specific substrates, for example, the enzymes used in desizing do not affect the cellulose, so there is no loss of cotton strength. Fifth, enzymes are easy to control because their activity depends on optimal conditions. Sixth, enzymes are biodegradable. At the end of the reaction where the enzymes were used, we can simply drain off the remaining solution because the enzymes are biodegradable and do not produce toxic waste in degradation, therefore no pollution.

Enzymes used in textile processing

enzymatic desire

Amylases are used to remove starch-based sizing for improved and uniform wet processing in the textile industry. An amylase enzyme can be used for desizing processes at low temperature (30-60ºC) and the optimum pH is 5.5-6.5. The advantage of these enzymes is that they are specific for starch, removing it without damaging the supporting tissue.

enzymatic pickling

Scouring is the removal of non-cellulosic material present on the cotton surface. In general, cellulase and pectinase are combined and used for bioscouring . In this case, pectinase destroys the cuticle structure of the cotton by digesting pectin and removing the connection between the cuticle and the cotton fiber body, while cellulase can destroy the cuticle structure by digesting the cellulose of cotton. the primary wall immediately below the cuticle of the cotton.

enzyme bleaching

The purpose of cotton bleaching is to decolorize the natural pigments and give the fibers a pure white appearance. Mainly flavonoids are responsible for the color of cotton. The most common industrial bleaching agent is hydrogen peroxide. Conventional cotton preparation requires large amounts of alkaline chemicals and consequently large amounts of rinse water are generated.

However, radical reactions of the bleaching agents with the fiber can lead to a decrease in the degree of polymerization and thus severe damage. Therefore, replacing hydrogen peroxide with an enzymatic bleaching system would not only lead to better product quality due to less fiber damage, but also substantial savings in the washing water needed to remove the peroxide. of hydrogen. An alternative to this process is to use a combination of suitable enzyme systems.

Amyloglucosidases, pectinases and glucose oxidases are selected that are compatible in terms of their active pH and temperature range.

biopolished

Biopolishing is a finishing process that improves the quality of the fabric by mainly reducing the fuzziness and pilling property of the cellulosic fiber. The goal of the process is the removal of cotton microfibrils through the action of the enzyme cellulase. The biopolishing treatment gives the fabric a cleaner surface, a fresher feel, shine and a softer feel.

EPISODE 2

anatomy of a cell

Cells are the basic and fundamental unit of life. So if we were to divide an organism at the cellular level, the smallest independent component we would find would be the cell. Cell Definition

"A cell is defined as the smallest basic unit of life that is responsible for all life processes."

Cells are the structural, functional and biological units of all living things. A cell can replicate itself independently. Therefore, they are known as the building blocks of life.

Each cell contains a fluid called the cytoplasm, which is enclosed by a membrane. Various biomolecules such as proteins, nucleic acids, and lipids are also present in the cytoplasm. Furthermore, cellular structures called cell organelles are suspended in the cytoplasm.

What is a cell?

A cell is the structural and fundamental unit of life. The study of cells from their basic structure to the functions of each cell organelle is called Cell Biology. Robert Hooke was the first biologist to discover cells.

All organisms are made up of cells. They can be made up of a single cell (unicellular) or many cells (multicellular). Mycoplasmas are the smallest cells known. Cells are the building blocks of all living things. They provide structure to the body and convert nutrients extracted from food into energy.

Cells are complex and their components perform various functions in an organism. They are of different shapes and sizes, much like the bricks in buildings. Our body is made up of cells of different shapes and sizes.

Cells are the lowest level of organization in all life forms. From one organism to another, the cell count can vary. Human beings have more number of cells compared to that of bacteria.

Cells comprise various cellular organelles that perform specialized functions to carry out life processes. Each organelle has a specific structure. The hereditary material of organisms is also present in cells.

cell discovery

The discovery of cells is one of the remarkable advances in the field of science. It helps us to know that all organisms are made up of cells, and these cells help carry out various life processes. The structure and functions of cells helped us to better understand life.

Who discovered cells?

Robert Hooke discovered the cell in 1665. Robert Hooke looked at a piece of bottle cork under a compound microscope and noticed tiny structures that reminded him of small rooms. Consequently, he called these "rooms" cells. However, his compound microscope had limited magnification and so he was unable to see any detail in the structure. Due to this limitation, Hooke concluded that they were non-living entities.

Later, Anton Van Leeuwenhoek viewed cells under another compound microscope at higher magnification. This time, he had noticed that the cells exhibited some kind of movement (motility). As a result, Leeuwenhoek concluded that these microscopic entities were "alive". Eventually, after a series of other observations, these entities were named animalcules.

In 1883, Robert Brown, a Scottish botanist, provided the first insights into cell structure. He was able to describe the nucleus present in orchid cells.

Cell characteristics

The following are the various essential characteristics of cells:

Cells provide structure and support to the body of an organism.

The interior of the cell is organized into different individual organelles surrounded by a separate membrane.

The nucleus (main organelle) contains the genetic information necessary for cell growth and reproduction.

Each cell has a nucleus and membrane-bound organelles in the **cytoplasm.**

Mitochondria, a double-membrane organelle, is primarily responsible for energy transactions vital to cell survival.

Lysosomes digest unwanted materials in the cell.

The endoplasmic reticulum plays an important role in the internal organization of the cell by synthesizing selective molecules and processing, directing, and sorting them into their proper locations. Cell structure

Cellular structure comprises individual components with specific functions essential for carrying out life processes. These components include: cell wall, cell membrane, cytoplasm, nucleus, and cell organelles. Read on to explore more information on cell structure and function.

Cellular membrane

The cell membrane supports and protects the cell. It controls the movement of substances in and out of cells. It separates the cell from the external environment. The cell membrane is present in all cells.

The cell membrane is the outer shell of a cell within which all other organelles, such as the cytoplasm and nucleus, are enclosed. It is also known as the plasma membrane.

By structure, it is a porous membrane (with pores) that allows the movement of selective substances in and out of the cell. In addition to this, the cell membrane also protects the cell component from damage and leakage.

It forms the wall-like structure between two cells, as well as between the cell and its environment.

Plants are immobile, so their cellular structures are well adapted to protect them from external factors. The cell wall helps to reinforce this function. Every cell in the body is enclosed by a cell (plasma) membrane. The cell membrane separates material from outside the cell, extracellular, from material inside the cell, intracellular. Maintains the integrity of a cell and controls the passage of materials in and out of the cell. All materials within a cell must have access to the cell membrane (the cell boundary) for necessary exchange.

The cell membrane is a double layer of phospholipid molecules. Proteins in the cell membrane provide structural support, form channels for the passage of materials, act as receptor sites, function as carrier molecules, and provide identification markers.

Cellular wall

The cell wall is the most prominent part of the plant cell structure. It is made up of cellulose, hemicellulose and pectin.

The cell wall is present exclusively in plant cells. Protects the plasma membrane and other cell components. The cell wall is also the outermost layer of plant cells.

It is a rigid and rigid structure that surrounds the cell membrane.

Provides shape and support to cells and protects them from shock and mechanical injury.

Cytoplasm

Cytoplasm is a thick, clear, jelly-like substance present within the cell membrane.

Most of the chemical reactions within a cell take place in this cytoplasm.

Cellular organelles such as the endoplasmic reticulum, vacuoles, mitochondria, ribosomes, are suspended in this cytoplasm.

Nucleus

The nucleus contains the hereditary material of the cell, DNA.

It sends signals to cells to grow, mature, divide, and die.

The nucleus is surrounded by the nuclear envelope that separates the DNA from the rest of the cell.

The nucleus protects the DNA and is an integral component of a plant's cell structure. nucleus and nucleolus

The nucleus, formed by a nuclear membrane around a fluid nucleoplasm, is the control center of the cell. Chromatin strands in the nucleus contain deoxyribonucleic acid (DNA), the cell's genetic material. The nucleolus is a dense region of ribonucleic acid (RNA) in the nucleus and is the site of ribosome formation. The nucleus determines how the cell will function , as well as the basic structure of that cell.

cytoplasmic organelles

Cytoplasmic organelles are "little organs" that are suspended in the cytoplasm of the cell. Each type of organelle has a defined structure and a specific role in the function of the cell. Examples of cytoplasmic organelles are mitochondria, ribosomes, endoplasmic reticulum, Golgi apparatus, and lysosomes.

Protein Construction.

Proteins are the end products of the decoding process that begins with the information in the cellular DNA. As the workhorses of the cell, proteins make up structural and motor elements in the cell, and serve as catalysts for virtually all biochemical reactions that occur in living things. This incredible variety of functions is derived from amazingly simple code that specifies an enormously diverse set of structures.

In fact, each gene in cellular DNA contains the code for a unique protein structure. These proteins not only assemble with different amino acid sequences, but are also held together by different bonds and fold into a variety of three-dimensional structures. The folded shape, or conformation, depends directly on the linear amino acid sequence of the protein.

What are proteins made of?

The building blocks of proteins are amino acids, which are small organic molecules consisting of an alpha (central) carbon atom bonded to an amino group, a carboxyl group, a hydrogen atom, and a variable component called a side chain (see more down) . Within a protein, multiple amino acids are linked together by peptide bonds, thus forming a long chain. Peptide bonds are formed by a biochemical reaction that removes a molecule of water when it joins the amino group of one amino acid with the carboxyl group of a neighboring amino acid. The linear sequence of amino acids within a protein is considered the primary structure of the protein.

Proteins are built from a set of just twenty amino acids, each of which has a unique side chain. Amino acid side chains have different chemistries. The largest group of amino acids have nonpolar side chains. Several other amino acids have positively or negatively charged side chains, while others have polar but uncharged side chains. The chemistry of amino acid side chains is critical to protein structure because these side chains can bind together to keep a length of protein in a certain shape or conformation. Charged amino acid side chains can form ionic bonds, and polar amino acids are capable of forming hydrogen bonds. Hydrophobic side chains interact with each other through weak van der Waals interactions. The vast majority of the bonds formed by these side chains are non-covalent . In fact, cysteines are the only amino acids capable of forming covalent bonds, which they do with their particular side chains. Due to side chain interactions, the sequence and location of the amino acids in a particular protein guide where the bends and folds occur in that protein. The primary structure of a protein, its amino acid sequence, drives the intramolecular folding and joining of the linear chain of amino acids, ultimately determining the protein's unique three-dimensional shape. Hydrogen bonds between amino groups and carboxyl groups in neighboring regions of the protein chain sometimes cause certain folding patterns. Known as alpha helices and beta sheets, these stable folding patterns make up the secondary structure of a protein. The final form assumed by a newly synthesized protein is usually the most favorable from an energetic point of view. As proteins fold, they try out a variety of conformations before reaching their final shape, which is single and compact. Folded proteins are stabilized by thousands of noncovalent bonds between amino acids. Additionally, chemical forces between a protein and its immediate environment contribute to the protein's shape and stability. For example, proteins that dissolve in the cell cytoplasm have hydrophilic (water-loving) chemical groups on their surfaces, while their hydrophobic (water-rejecting) elements tend to be tucked inside. In contrast, proteins that are inserted into cell membranes show some hydrophobic chemical groups on their surface, specifically in those regions where the protein surface is exposed to membrane lipids. However, it is important to note that fully folded proteins do not freeze into shape. Rather, the atoms within these proteins are still capable of small movements.

Despite their macromolecular state, proteins are still too small to be seen, even under a microscope. This forces scientists to employ indirect approaches to learn about their structure and shape. X-ray crystallography is the standard technique for analyzing protein structures. Crystals of purified protein are placed under a beam of X-rays, and the pattern of deflected X-rays is used to calculate the locations of thousands of atoms within the crystal.

What causes proteins to take their final forms?

Once their amino acid building blocks are chained together, proteins can theoretically fold into their mature shapes without much effort. However, the cytoplasm actually contains numerous additional macromolecules that can interact with a protein that is only folded in half. Proteins can form large aggregates in cells if they create inappropriate connections with neighboring proteins, preventing them from folding properly. Cells therefore rely on proteins known as chaperones to prevent proteins from binding to unwanted folding partners.

The role of a chaperone protein is to enclose and protect a target protein from degradation as it folds. Multiple molecules of the bacterial chaperone GroEL, for example, create a vacuum around unfolding proteins. The chamber is covered by a lid made of

GroES molecules , which function as a second carbine. Although all chaperone proteins in eukaryotes serve the same general purpose, they come from different families.

There is a plethora of cellular chaperone proteins. When they bind and unbind polypeptides, these chaperones use the energy of ATP. Protein refolding is another process that chaperones aid in throughout cellular metabolism. Proteins in their folded state are extremely unstable structures that can be denatured or unfolded with little effort. Although proteins are held together by thousands upon thousands of bonds, the vast majority of these bonds are noncovalent. and relatively weak. Unfolded proteins make up a small percentage of all cellular proteins even under normal conditions. An increase of even a few degrees in core body temperature can have a dramatic effect on the rate of development. For situations like this, using chaperones to repair the broken proteins instead of starting from scratch is the most efficient course of action. Interestingly, "heat shock" causes cells to produce more chaperone proteins

In other words, what do protein families consist of?

The specific function of a protein is determined by the way its exposed surfaces interact with other molecules; This is because all proteins bind to other molecules to carry out their tasks. Proteins that share a similar general structure have a higher chance of interacting with the same set of molecules. There is a strong correlation between the functions performed by proteins of the same family.

Long regions of conserved amino acid sequences are another hallmark of proteins within the same family. These sequences have been highly conserved throughout evolution because they are essential for the catalytic activity of the protein. When it comes to interacting with common intracellular signaling proteins , for example, the amino acid sequences in cellular receptor proteins tend to be more similar than in their binding sites, which receive chemical signals from outside the cell. Multiple members of a protein family are possible and presumably originated from early gene duplications. Over time, these copies led to adjustments in protein functionality, which in turn expanded the range of capabilities available to organisms.

The structure of proteins.

There can be up to four different structural orders in a single protein. Proteins have a primary structure formed by the chain of amino acids. Certain portions of the polypeptide chain fold into stable patterns due to interactions between the various amino acids. Alpha helices and beta sheets are two types of secondary structure. Tertiary structure is the result of interactions between secondary structures.

Additionally, there are scenarios where a protein requires the assistance of multiple copies to complete an operation. Specifically, in this context, each protein is referred to as a subunit. The quaternary structure of a protein represents its final fully assembled state, consisting of all of its individual subunits.

Chapter 3

Metabolism

The term "metabolism" is used to refer to the sum total of cellular and organic chemical reactions necessary to sustain life. To simplify things, we can classify metabolism into two large fields:

To obtain energy by breaking down molecules, a process known as catabolism is carried out.

Synthesis of all the molecules required by cells; also known as anabolism.

Nutrition and nutrient availability play an important role in metabolism. Bioenergetics refers to the biochemical and metabolic processes through which a cell generates energy. The creation of energy is a crucial part of the metabolic process. Energy, metabolism and nutrition

Metabolic health begins with proper nutrition. Metabolic pathways require nutrients for the breakdown of which generate energy. In turn, the body needs this energy to create compounds such as proteins and nucleic acids (DNA, RNA).

When talking about nutrients and metabolism, it's important to consider things like the body's demand for certain substances, how those substances are used within the body, how much is required, and what happens to health if that amount isn't met.

Calories (from essential nutrients) provide fuel, while essential nutrients provide substances that the body needs but cannot make. A healthy diet provides the nutrients necessary for growth, maintenance and repair of tissues, as well as for optimal physiological performance.

carbon, hydrogen, oxygen, nitrogen, phosphorus, sulfur and more than 20 other inorganic elements. Carbohydrates, fats and proteins contribute to the main nutrients. Vitamins, minerals and water are also important. What is the function of metabolism, exactly?

- Enzymes are used by the digestive system to:
- digest proteins by separating their constituent amino acids
- create fatty acids from fats
- convert complex carbohydrates to glucose (for example, glucose)

Do you need energy? The body can use sugar, amino acids and fatty acids. The chemicals are absorbed into the blood and delivered to the cells.

Other enzymes speed up or regulate the chemical reactions involved in the " metabolism " of these substances once they have entered cells. These reactions allow the body to use the energy contained in these chemicals or store it in various tissues such as the liver, muscle, and fat.

In metabolism, two types of processes constantly work in tandem:

restore strength and restore tissues (called anabolism)

to obtain additional fuel for human operations by breaking down biological tissues and storing energy (called catabolism)

"Anabolism" (uh-NAB-uh- liz-um) stands for "building metabolism" and focuses on building and storing molecules. It helps the body make new cells, keep old cells healthy, and store energy for later use. Anabolism is the process by which simple molecules such as glucose, proteins and fats are transformed into larger and more complex ones.

All cellular activity is based on a process called catabolism (kuh -TAB- uh- liz -um), often known as destructive metabolism. Energy is released when cells break down large molecules (mainly carbohydrates and lipids). This generates heat for the body, powers muscle contractions, and gets the blood pumping.

Waste products are excreted from the body through the skin, kidneys, lungs, and intestines as the body breaks down complex chemical units into simpler compounds.

What factors regulate metabolism?

Various endocrine hormones regulate metabolic rate and metabolic expenditure. The hormone thyroxine is secreted by the thyroid gland and plays an important role in regulating the rate at which the body's metabolism operates.

Anabolic processes, for example, tend to increase after a meal. That's because food raises blood glucose levels, which the body uses for almost all of its energy needs. When the pancreas senses a rise in blood sugar, it secretes the hormone insulin, which tells cells to increase their repair processes.

The metabolic process involves a wide variety of chemical reactions. As a result, it's no surprise that many people see it in its most basic sense: as a factor in how quickly our bodies gain or lose excess weight. Calories provide this function. The amount of usable energy in a given food is expressed as a number of calories. Compared to an apple, the number of calories in a chocolate bar is substantially higher, which means that it provides more energy to the body. Calories are stored in the body, primarily as fat, much like gasoline is stored in a car's gas tank until it's needed to fuel the engine. Gasoline may spill onto the road if a car's tank is too full. Likewise, excess calories "spill over" as additional body fat if a person consumes an excessive amount of food.

A person's daily caloric expenditure is determined by factors such as exercise level, body composition (particularly the ratio of fat to muscle), and resting metabolic rate (BMR). A person's basal metabolic rate (BMR) is the number of calories their body "burns" while at rest.

A person's propensity to gain weight can be affected by their basal metabolic rate. A person with a low basal metabolic rate (who, as a result, burns fewer calories while resting or sleeping), would, on average, gain more weight than someone of comparable size with an average BMR, even if they both consume the same number of calories and perform the same amount of physical activity.

Both genetics and health problems can influence BMR. People with higher BMRs tend to be those with more muscle and less fat on their bodies. However, there are techniques for people to modify their BMR. A person's basal metabolic rate (BMR) can be increased by a variety of means, including exercise.

You may have heard the phrase "slow metabolism" used in reference to people's inability to maintain a healthy weight, but what exactly does it mean? Can we really blame the metabolism? Is this true? If so, can your metabolism be boosted to allow you to burn more calories?

There is a correlation between your metabolism and your weight. **But** contrary to conventional wisdom, a slow metabolism is rarely the cause of excessive weight gain.

Although your metabolism determines your body's basic energy needs, the amount you eat and drink along with the amount of physical exercise you engage in are the elements that ultimately decide your weight.

Metabolism:

Convert food into energy

Metabolism is the process by which your body converts what you eat and drink into energy. During this intricate process, calories from food and beverages mix with oxygen to release the energy your body needs to function.

Even when you're at rest, your body needs energy for all of its "hidden" tasks, like breathing, circulating blood, changing hormone levels, and growing and repairing cells. The number of calories your body consumes to carry out these basic tasks is known as your basal metabolic rate, which you can call metabolism.

Several factors determine your particular basal metabolism, including:

The size and composition of your body. Larger or more muscular people burn more calories, even at rest.

your gender Men typically have less body fat and more muscle than women of the same age and weight, which means men burn more calories.

Your age. As you age, the amount of muscle tends to decrease and fat makes up more of your weight, slowing down calorie burning.

The energy required for the essential operations of your body remains fairly stable and is not easily adjusted.

In addition to your basal metabolic rate, two other factors affect how many calories your body burns each day:

Food processing (thermogenesis) (thermogenesis). Digesting, absorbing, transferring, and storing the food you eat also requires energy. About 10 percent of the calories from carbohydrates and protein you eat are used during the digestion and absorption of food and nutrients.

physical activity Physical activity and exercise, such as playing tennis, walking to the store, chasing the dog, and any other movement, account for the rest of the calories your body burns each day. Physical activity is by far the most variable of the factors that affect how many calories you burn each day.

Scientists call the activity you do all day that is not deliberate exercise thermogenesis (NEAT) (NEAT). This activity involves pacing from room to room, activities like gardening, and even fidgeting. NEAT represents approximately 100 to 800 calories used every day .

metabolism and weight

It can be tempting to blame your metabolism for weight gain. But because metabolism is a natural process, your body has various mechanisms that regulate it to meet your specific demands.

Only in rare situations do you have substantial weight gain due to a medical condition that slows your metabolism, such as Cushing's syndrome or having an underactive (hypothyroid) thyroid gland (hypothyroidism).

Unfortunately, weight gain is a difficult process. It is likely to be a combination of genetic makeup, hormonal regulation, diet composition, and the impact of the environment on your lifestyle, including sleep, physical activity, and stress.

All of these causes result in an imbalance in the energy equation. You gain weight when you eat more calories than you burn, or when you burn fewer calories than you eat.

While it's true that some people seem to be able to lose weight more quickly and easily than others, everyone loses weight when they burn more calories than they eat. To lose weight, you must create an energy deficit by eating fewer calories or by increasing the number of calories you burn through physical activity, or both.

A closer look at physical activity and metabolism

While you don't have much control over the speed of your basal metabolism, you can regulate the number of calories you burn based on how active you are. The more active you are, the more calories you will burn. In fact, some people who are thought to have a fast metabolism are likely to be more active, and perhaps more restless, than others.

Aerobic exercise is the most efficient way to burn calories and includes activities like walking, biking, and swimming. **As a general goal, add at least 30 minutes of physical activity to your daily routine.**

If you want to lose weight or reach certain fitness goals, you may need to increase the time you spend exercising even more. If you can't carve out time for a longer workout, try 10-minute action pieces throughout the day. Remember, the more active you are, the greater the benefits.

Experts also encourage strength training routines, such as lifting weights, at least twice a week. Strength training is vital because it helps build muscle. Muscle tissue consumes more calories than fat tissue.

Any extra movement helps burn calories. Find ways to walk and move a few minutes more each day than the day before. Taking the stairs more often and parking farther away at the store are simple strategies for burning extra calories. Even hobbies like gardening, washing the car, and doing housework burn calories and help you lose weight.

no miracle bullet

Don't look for dietary supplements to help burn calories or lose weight. Products that claim to speed up your metabolism are usually more hype than help, and some can create unwanted or even serious side effects.

Manufacturers of dietary supplements are not required by the US Food and Drug Administration to establish that their products are safe or effective, so view these products with caution. Always tell your doctors about any supplements you take.

There is no easy way to lose weight. The foundation for weight loss still needs to be built on physical activity and diet. Eat fewer calories than you expend and you will lose weight.

The Dietary Guidelines for Americans suggest cutting calories by 500 to 700 calories per day to lose 1 to 1.5 pounds (0.5 to 0.7 kilograms) per week . If you can add some physical exercise to your day, you'll reach your weight loss goals even faster.

Photosynthesis

Photosynthesis is the process by which plants use sunlight, water, and carbon dioxide to produce oxygen and energy in the form of sugar. It would be impossible to overstate the importance of photosynthesis in sustaining life on Earth. If photosynthesis stopped, there would soon be little food or other biological material on Earth. Most life would die out, and in time the Earth's atmosphere would become almost barren of gaseous oxygen. The only species capable of thriving under such conditions would be chemosynthetic bacteria, which can use chemical energy from specific inorganic chemicals and are therefore not dependent on light energy conversion.

The energy produced by photosynthesis carried out by plants millions of years ago is responsible for the fossil fuels (ie coal, oil, and gas) that power industrial society. In past periods, green plants and microscopic plant-eating organisms spread faster than they were consumed, and their remains were deposited in the Earth's crust through sedimentation and other geological processes. There, sheltered from oxidation, these organic residues were slowly transformed into fossil fuels. These fuels not only provide much of the energy used in businesses, homes, and transportation, but also serve as raw materials for plastics and other synthetic items. Unfortunately, modern society is consuming in a few centuries the excess photosynthetic productivity acquired over millions of years. Consequently, the carbon dioxide that has been removed from the air to produce carbohydrates in photosynthesis for millions of years is being returned at an incredibly fast rate. The concentration of carbon dioxide in Earth's atmosphere is increasing faster than ever in Earth's history, and this phenomenon is likely to have severe effects on Earth's climate.

The requirements for food, materials, and energy in a world where the human population is rapidly expanding have led to the need to improve both the amount of photosynthesis and the efficiency of converting photosynthetic output into useful goods for people. One response to such needs, the so-called Green Revolution, launched in the mid-20th century, achieved substantial gains in agricultural productivity through the use of chemical fertilizers , insect and plant disease management, plant breeding and mechanical tillage, harvesting, and crop processing. This effort reduced severe famines in a few places in the world despite rapid population expansion, but it did not eliminate widespread hunger. Furthermore, beginning in the early 1990s, the rate at which yields of major crops increased began to slow. This was especially true for rice in Asia. The rising costs associated with maintaining high rates of agricultural production, which required ever-increasing inputs of fertilizers and pesticides and the continual development of new types of plants, also made it difficult for farmers in many countries.

The biochemistry of metabolism.

The main organic components present in a biological organism, and therefore the most crucial for metabolism, are carbohydrates, proteins and lipids (fats). In addition, cells contain large amounts of water, salts, and minerals.

Furthermore, nucleic acids and other chemical residues are crucial elements of cells. Organic molecules consist of carbon, hydrogen, oxygen, phosphorus, and sulfur components.

The metabolic process is based on proper nutrition. Collagen, insulin, antibodies, enzymes, and hemoglobin are examples of proteins that provide important structural and metabolic functions in the body. Energy storage, respiration, and muscle contraction depend on the metabolism of carbohydrates, which include starches, cellulose, and sugars. The manufacture of hormones and organelles, as well as cell transport, signaling, and heat production, all depend on lipids.

Dietary nutrients provide the chemicals necessary for cellular processes to function smoothly, making them an essential factor in metabolism. Water, oxygen, minerals, and other inorganic molecules are also necessary for metabolic reactions, in addition to necessary nutrients such as carbohydrates and proteins. Proteins are made up of amino acids, which are predominantly generated within the body. However, some, such as lysine, phenylalanine, valine , and histidine , are inaccessible to mammals and must be obtained through the diet.

Vitamins play a crucial role in the synthesis of enzymes and, by extension, in the proper functioning of the metabolic pathways they control. Coenzyme vitamins, including niacin, thiamin and others, are essential for proper body function and their lack can have devastating effects.

Metabolism and Hormones.

Hormones control the body's metabolism. Insulin, glucose, adrenaline, and hormones that regulate the thyroid fall into this category. Different organs and tissues create hormones with different metabolic effects. Gene expression regulates the generation of hormones.

The pancreas secretes insulin, which plays a key role in controlling how cells use glucose for fuel. Like insulin, glucagon prevents blood glucose levels from falling to life-threatening levels. During times of high stress, the hormone epinephrine causes the body to produce more glucose. In the liver, it increases glucose production. The activity of the liver's gluconeogenesis-suppressing enzyme, hepatic glycogenesis, is influenced by thyroid hormones, which also affect cholesterol and carbohydrate metabolism. Thyroid hormone activity must be normal to provide adequate growth-related metabolism.

Factors That May Have an Impact on Your Metabolism

A person's health can be affected if their metabolism is disturbed in any way. A common condition in which insulin production fails to control blood glucose levels is diabetes.

An increased metabolic rate associated with hyperthyroidism can cause irregular heartbeat, altered menstrual cycles, weight loss, and sweating. Rapid weight gain, infertility, and joint pain are just some of the symptoms of hypothyroidism, a disorder of the thyroid gland. Women are more likely to experience symptoms of hyperthyroidism or hypothyroidism. Metabolic disorders vary in severity and impact on a patient's quality of life, but all require medical attention. Some of these diseases are directly related to the way of life that one leads, while others are hereditary.

Chapter 4

enzymes

Many chemical reactions in the body can be sped up with the help of enzymes. Its thousands of functions, including respiration, digestion, muscle and nerve function, would not be possible without them.

There are thousands of enzymes in every human cell. Enzymes help catalyze chemical processes within each cell.

Since the enzymes are not broken down during the process, the cell can recycle them. To keep the body running and healthy, enzymes help in specific processes that are essential.

Protein catalysts known as enzymes play an important role in the body's ability to speed up chemical reactions. Enzymes play a critical role in many bodily processes, including digestion, liver function, and many others. The body cannot handle either too much or too little of an enzyme, and either can have negative effects. Blood enzyme tests are another tool used by doctors to diagnose illness and injury.

Enzymes: what are they?

Proteins called enzymes increase the rate at which our bodies metabolize food and other substances. Some substances build up while others break down. Enzymes are present in all living things.

Enzymes are chemical compounds that our body generates on its own. however , enzymes can also be found in foods and manufactured products.

Why are they so important?

Enzymes play numerous roles, but one of the most crucial is helping with digestion. Simply put, digestion is the transformation of nutrients from food into a usable form. Saliva, pancreatic secretions, the digestive tract, and the stomach contain enzymes. They metabolize proteins , carbohydrates and lipids. These substances are fuel for enzymes, which help in cell growth and repair.

In addition, enzymes help in

- Breathing.

- Muscle building.
- The functioning of the nervous system.
- purging our systems of harmful substances.

Exactly how many different forms of enzymes are there ?

Every human body contains thousands of unique enzymes. There is only one function for each type of enzyme. One of those enzymes is sucrase , which is responsible for the digestion of sucrose. The milk sugar called lactose can be digested with the help of an enzyme called lactase.

Here are some of the most commonly used digestive enzymes:

- Carbohydrates are converted into sugars by the enzyme carbohydrase .
- Lipids are hydrolyzed by lipase into fatty acids.
- Protease is a proteinase that cleaves a protein chain into its constituent amino acids.

DIFFERENT COMPONENTS OF ENZYMES

When you think of an enzyme, what do you imagine?

The "active site" of each enzyme has a different shape. A substrate is the material that is metabolized by an enzyme . There is also a special shape for the substrate. For the enzyme to work, it needs a substrate that can interact with it.

The effects of heat and acidity on enzymes.

When it comes to enzymes, the settings must be correct. Enzymes can undergo conformational changes if conditions are not optimal. So they are not compatible with the substrates and do not work as intended.

Each enzyme has a specific pH and temperature sweet spot:

Enzymes can be damaged by extreme changes in pH. You can't have an acidic or basic environment with them because they won't work properly. Pepsin, an enzyme found in the stomach, is an example of an enzyme that catalyzes this process. The effectiveness of pepsin is reduced if stomach acid levels are too low.

The optimum temperature for enzyme activity is around 37 degrees Celsius (98.6 degrees Fahrenheit). Protein synthesis and other biochemical reactions speed up in a warmer environment. However, the enzyme will no longer work if the temperature is too high. As a result, a high temperature can interfere with normal bodily processes.

COMMON HEALTH PROBLEMS
How do enzyme deficiencies affect human health?

The lack of a certain enzyme is a common cause of metabolic disorders. They can be inherited from one generation to the next (inherited). Here are some examples of metabolic disorders that run in families:

For those with Fabry disease , the inability to produce fat-digesting enzymes (alpha- galactosidase A) is a major problem (lipids).

Enzymes involved in the production of myelin, a protective covering for nerve cells, are disrupted in Krabbe disease (globoid cell leukodystrophy) (Central Nervous System).

The enzymes responsible for metabolizing certain branched-chain amino acids are altered in patients with maple syrup urine disease.

In addition to the above, enzyme imbalances have been linked to the following medical problems:

In Crohn's disease , an autoimmune response from the gut can be influenced by an imbalance of the bacteria in your gut (gut microbiome). This can influence the severity and course of Crohn 's disease .

Insufficient production of digestive enzymes by the pancreas is known as exocrine pancreatic insufficiency (EPI). Inability to digest food and absorb its nutrients. PID can develop as a result of chronic pancreatitis, pancreatic cancer, diabetes, or cystic fibrosis.

Lack of lactase, an enzyme needed to digest sugar found in milk and other dairy products, causes lactose intolerance.

What are the diagnostic applications of enzyme tests?

There are many different enzyme and protein blood tests that your doctor can use to diagnose disease. Elevated levels of some liver enzymes, for example, can indicate liver disease.

PROPER ENZYME CARE

Should I take enzymes?

Enzymes can usually be obtained through a balanced diet for people without ongoing medical problems. However, your doctor may recommend that you take enzyme supplements if you have a specific medical condition. A common practice among PID sufferers is to take a digestive enzyme before each meal. In turn, this improves your digestion and absorption of nutrients from food. Before taking any enzyme supplement, you should consult with your doctor.

: Can Medications Have an Effect?

Enzyme inhibitors are sometimes prescribed as a treatment for other conditions. Some enzymes depend on bacteria for optimal function, but antibiotics can kill those bacteria. And that's why antibiotics can upset your stomach. However, the good bacteria that are essential for digestion are also killed in the process of eliminating the bacteria that cause your illness.

Taking statins (cholesterol-lowering drugs) can increase the levels of enzymes in the liver and muscles. They may be able to make the body more susceptible to liver and muscle damage.

WHO SHOULD I CALL?

How soon should I seek medical attention if I suspect an enzyme problem?

If you suspect that you have an enzyme disorder, a doctor will want to take a blood sample. Those who experience any of the following problems should see a doctor immediately:

Symptoms of abdominal discomfort.

- Gas or bloating.
- Diarrhea.
- Fatigue.
- Feeling sick to your stomach and vomiting.
- Sudden and disconcerting weight loss.
- Having a low number of red blood cells (anaemia).
- Internal bleeding from the stomach or intestines.

Why are they so important?

Throughout the body, enzymes help in crucial ways. I give you some illustrations:

Digestive System: **Enzymes help break down larger, more complex molecules into simpler ones, such as glucose, which the body can use for fuel.**

DNA is contained in every cell in the body and can replicate. When a cell divides, it must make a copy of its DNA. The DNA coils unwind with the help of enzymes.

Toxins in the body are neutralized by enzymes produced by the liver. To achieve this, it makes use of a series of enzymes that help in the breakdown of toxins.

Everything is in place to make this a successful company.

Enzymes require specific environments to function properly. At a temperature of about 37 degrees Celsius (98.6 degrees Fahrenheit), most of the enzymes in the human body function at peak efficiency. They could still work at lower temperatures, albeit much more slowly.

The enzyme undergoes a conformational change when exposed to extreme temperatures or pH, rendering the active site unsuitable for substrate binding. The process is denaturing.

Enzymes have a range of tolerance to acidity. Enzymes in the intestines, for example, thrive at a pH of 8 while those in the stomach thrive at a pH of 1.5 due to the much more acidic environment of the stomach.

Cofactors

To carry out their biological tasks, certain enzymes require the presence of other molecules, known as cofactors. The term "cofactor" can refer to an ion or a coenzyme, both of which are essential to the biochemical process.

The ions, which are inorganic molecules, bind loosely to the enzyme to keep it active. Coenzymes, on the other hand, are organic molecules that form weak bonds with enzymes and make their work easier.

The term "prosthetic group" is used to describe the chemical bond formed between a cofactor and an enzyme.

Inhibition

When maintaining proper function of body systems, it is sometimes necessary to reduce enzyme activity. If an enzyme produces too much of a product, for example, the body must find a way to slow or stop the enzyme's activity.

Several factors can limit enzyme activity levels, including:

Competitive Inhibitors – **This inhibitory molecule blocks the active site so that the substrate has to compete with the inhibitor to bind to the enzyme.**

Non-competitive inhibitors – **This molecule binds to an enzyme** somewhere other than the active site and reduces how effectively it works.

Non-competitive inhibitors: This inhibitor binds to the enzyme and the substrate. Products leave the active site less easily, slowing down the reaction.

Irreversible inhibitors: This is an irreversible inhibitor, which binds to an enzyme and permanently inactivates it.

Examples of Specific Enzymes

There are thousands of enzymes in the human body to perform around 5,000 different functions. Some examples include:

Lipases: This group of enzymes helps digest fats in the intestine.

Amylase: In saliva, amylase helps convert starches into sugars.

Maltase: This also occurs in saliva and breaks down the sugar maltose into glucose.

Trypsin: These enzymes break down proteins into amino acids in the small intestine.

Lactase: Lactase breaks down lactose, the sugar in milk, into glucose and galactose.

Acetylcholinesterase – These enzymes break down the neurotransmitter acetylcholine in nerves and muscles.

Helicase: Helicase enzymes unwind DNA.

THE END

www.ingramcontent.com/pod-product-compliance
Lightning Source LLC
Chambersburg PA
CBHW052355220526
45465CB00003BA/1110